Marx Joyce
Hardy Austen
Defoe Abbott Melville Montaigne Chesterton Hugo Grimm
Machiavelli Cooper Emerson Eliot
Haggard
Stoker Christie Molière
Wilde Carroll Maupassant Byron
Garnett Engels Schiller
Goethe Fitzgerald Hawthorne Smith Kafka
Cotton Einstein Dostoyevsky Hall
Baum Kipling Doyle Willis
Leslie Dumas Henry Nietzsche
Flaubert Turgenev Balzac
Stockton Vatsyayana Crane
Burroughs Verne
Curtis Tocqueville Vinci
Homer Widger Tolstoy Gogol Busch
Darwin Thoreau Whitman
Potter Freud Zola Twain Scott
Kant Jowett Lawrence Plato Harte
Stevenson Dickens
Andersen Hesse
London Descartes Cervantes Burton
Poe Aristotle Wells Voltaire
Hale James Hastings Cooke
Bunner Shakespeare Irving
Richter Chambers
Doré da Benedict Alcott
Dante Shaw Wodehouse
Swift Chekhov Pushkin
Newton

tredition®

tredition was established in 2006 by Sandra Latusseck and Soenke Schulz. Based in Hamburg, Germany, tredition offers publishing solutions to authors and publishing houses, combined with worldwide distribution of printed and digital book content. tredition is uniquely positioned to enable authors and publishing houses to create books on their own terms and without conventional manufacturing risks.

For more information please visit: www.tredition.com

TREDITION CLASSICS

This book is part of the TREDITION CLASSICS series. The creators of this series are united by passion for literature and driven by the intention of making all public domain books available in printed format again - worldwide. Most TREDITION CLASSICS titles have been out of print and off the bookstore shelves for decades. At tredition we believe that a great book never goes out of style and that its value is eternal. Several mostly non-profit literature projects provide content to tredition. To support their good work, tredition donates a portion of the proceeds from each sold copy. As a reader of a TREDITION CLASSICS book, you support our mission to save many of the amazing works of world literature from oblivion. See all available books at www.tredition.com.

 Project Gutenberg

The content for this book has been graciously provided by Project Gutenberg. Project Gutenberg is a non-profit organization founded by Michael Hart in 1971 at the University of Illinois. The mission of Project Gutenberg is simple: To encourage the creation and distribution of eBooks. Project Gutenberg is the first and largest collection of public domain eBooks.

Canned Fruit, Preserves, and Jellies: Household Methods of Preparation U.S. Department of Agriculture Farmers' Bulletin No. 203

Maria Parloa

Imprint

This book is part of TREDITION CLASSICS

Author: Maria Parloa
Cover design: Buchgut, Berlin – Germany

Publisher: tredition GmbH, Hamburg - Germany
ISBN: 978-3-8472-1294-2

www.tredition.com
www.tredition.de

Copyright:
The content of this book is sourced from the public domain.

The intention of the TREDITION CLASSICS series is to make world literature in the public domain available in printed format. Literary enthusiasts and organizations, such as Project Gutenberg, worldwide have scanned and digitally edited the original texts. tredition has subsequently formatted and redesigned the content into a modern reading layout. Therefore, we cannot guarantee the exact reproduction of the original format of a particular historic edition. Please also note that no modifications have been made to the spelling, therefore it may differ from the orthography used today.

U. S. DEPARTMENT OF AGRICULTURE.

FARMERS' BULLETIN No. 203.

Canned Fruit, Preserves, and Jellies:
HOUSEHOLD METHODS OF PREPARATION.
BY
MARIA PARLOA.

PREPARED UNDER THE SUPERVISION OF THE OFFICE OF
EXPERIMENT STATIONS,
A. C. TRUE, DIRECTOR.

WASHINGTON:
GOVERNMENT PRINTING OFFICE.
1917.

[3]

CANNING AND PRESERVING FRUIT.

INTRODUCTION.

The common fruits, because of their low nutritive value, are not, as a rule, estimated at their real worth as food. Fruit has great dietetic value and should be used generously and wisely, both fresh and cooked. Fruits supply a variety of flavors, sugar, acids, and a necessary waste or bulky material for aiding in intestinal movement. They are generally rich in potash and soda salts and other minerals. Most fresh fruits are cooling and refreshing. The vegetable acids have a solvent power on the nutrients and are an aid to digestion when not taken in excess.

Fruit and fruit juices keep the blood in a healthy condition when the supply of fresh meat, fish, and vegetables is limited and salt or smoked meats constitute the chief elements of diet. Fresh fruit is generally more appetizing and refreshing than cooked. For this reason it is often eaten in too large quantities, and frequently when underripe or overripe; but when of good quality and eaten in moderate quantities it promotes healthy intestinal action and rarely hurts anyone.

If eaten immoderately, uncooked fruit is apt to induce intestinal disturbances. If eaten unripe, it often causes stomach and intestinal irritation; overripe, it has a tendency to ferment in the alimentary canal. Cooking changes the character and flavor of fruit, and while the product is not so cooling and refreshing as in the raw state, it can, as a rule, be eaten with less danger of causing stomach or intestinal trouble. If sugar be added to the cooked fruit, the nutritive value will be increased. A large quantity of sugar spoils the flavor of the fruit and is likely to make it less easily digested.

Nowhere is there greater need of a generous supply of fruit than on the farm, where the diet is apt to be restricted in variety because of the distance from markets. Every farmer should raise a generous supply of the kinds of fruit that can be grown in his locality. Wives and daughters on the farms should find pleasure in serving these fruits in the most healthful and tempting form. There are a large number of simple, dainty desserts that can be prepared with fruit and without much labor. Such desserts should leave the pie as an occasional luxury instead of allowing it to be considered a daily necessity.

[4] In the season when each kind of fruit is plentiful and at its best a generous supply should be canned for the season when both fruit and fresh vegetables are scarce. A great deal of the fruit should be canned with little or no sugar, that it may be as nearly as possible in the condition of fresh fruit. This is the best condition for cooking purposes. A supply of glass jars does cost something, but that item of expense should be charged to future years, as with proper care the breaking of a jar need be a rare occurrence. If there be an abundance of grapes and small, juicy fruits, plenty of juice should be canned or bottled for refreshing drinks throughout the year. Remember that the fruit and juice are not luxuries, but an addition to the dietary that will mean better health for the members of the family and greater economy in the cost of the table.

FRESH AND PRESERVED FRUIT FOR THE MARKET.

If the supply of fruit is greater than the family needs, it may be made a source of income by sending the fresh fruit to the market, if there is one near enough, or by preserving, canning, and making jelly for sale. To make such an enterprise a success the fruit and work must be first class. There is magic in the word "Homemade," when the product appeals to the eye and the palate; but many careless and incompetent people have found to their sorrow that this word has not magic enough to float inferior goods on the market. As a rule large canning and preserving establishments are clean and have the best appliances, and they employ chemists and skilled labor. The home product must be very good to compete with the attractive goods that are sent out from such establishments. Yet for first-class homemade products there is a market in all large cities. All first-class grocers have customers who purchase such goods.

To secure a market get the names of several first-class grocers in some of the large towns. Write to them asking if they would be willing to try a sample of your goods. If the answer is favorable, send samples of the articles you wish to sell. In the box with the fruit inclose a list of the articles sent and the price. Write your name and address clearly. Mail a note and a duplicate list at the time you send the box.

Fixing the price of the goods is important. Make it high enough to cover all expenses and give you a fair return for your labor. The expenses will be the fruit, sugar, fuel, jars, glasses, boxes, packing material, wear and tear of utensils, etc., transportation, and commission. The commission will probably be 20 per cent of the selling price. It may be that a merchant will find that your prices are too high or too low for his trade, or he may wish to purchase the goods outright. In [5] any case it is essential that you estimate the full cost of the product and the value that you place on your labor. You will then be in a position to decide if the prices offered will compensate you for the labor and expense. Do not be tempted, for the sake of a little money, to deprive your family of the fruit necessary to health and pleasure.

PACKING AND SHIPPING.

Each jar or jelly glass must be wrapped in several thicknesses of soft paper (newspapers will answer). Make pads of excelsior or hay by spreading a thick layer between the folds of newspapers. Line the bottom and sides of the box with these pads. Pack the fruit in the padded box. Fill all the spaces between the jars with the packing material. If the box is deep and a second layer of fruit is to go in, put thick pasteboard or thin boards over the first layer and set the wrapped jars on this. Fill all the spaces and cover the top with the packing material. Nail on the cover and mark clearly: GLASS. THIS SIDE UP.

The great secret in packing is to fill every particle of space so that nothing can move.

PRINCIPLES OF CANNING AND PRESERVING.

In the preservation of foods by canning, preserving, etc., the most essential things in the processes are the sterilization of the food and all the utensils and the sealing of the sterilized food to exclude all germs.

BACTERIA, YEASTS, AND FERMENTATION.

Over one hundred years ago François Appert was the first to make practical application of the method of preserving food by putting it in cans or bottles, which he hermetically sealed. He then put the full bottles or cans in water and boiled them for more or less time, depending upon the kinds of food.

In Appert's time and, indeed, until recent years it was generally thought that the oxygen of the air caused the decomposition of food. Appert's theory was that the things essential to the preservation of food in this manner were the exclusion of air and the application of gentle heat, as in the water bath, which caused a fusion of the principal constituents and ferments in such a manner that the power of the ferments was destroyed.

The investigations of scientists, particularly of Pasteur, have shown that it is not the oxygen of the air which causes fermentation and putrefaction, but bacteria and other microscopic organisms.

[6] Appert's theory as to the cause of the spoiling of food was incorrect, but his method of preserving it by sealing and cooking was correct, and the world owes him a debt of gratitude.

In their investigations scientists have found that if food is perfectly sterilized and the opening of the jar or bottle plugged with sterilized cotton, food will not ferment, for the bacteria and yeasts to which such changes are due can not pass through the cotton. This method can not be conveniently followed with large jars.

Bacteria and yeasts exist in the air, in the soil, and on all vegetable and animal substances, and even in the living body, but although of such universal occurrence, the true knowledge of their nature and economic importance has only been gained during the last forty years.

There are a great many kinds of these micro-organisms. Some do great harm, but it is thought that the greater part of them are beneficial rather than injurious.

Bacteria are one-celled and so small they can only be seen by aid of a microscope. The process of reproduction is simple and rapid. The bacterium becomes constricted, divides, and finally there are two cells instead of one. Under favorable conditions each cell divides, and so rapid is the work that it has been estimated that one bacterium may give rise, within twenty-four hours, to seventeen millions of similar organisms. The favorable conditions for growth are moisture, warmth, and proper food.

Yeasts, which are also one-celled organisms, grow less rapidly. A bud develops, breaks off, and forms a new yeast plant. Some yeasts and some kinds of bacteria produce spores. Spores, like the dried seeds of plants, may retain their vitality for a long time, even when exposed to conditions which kill the parent organism.

Yeasts and nearly all bacteria require oxygen, but there are species of the latter that seem to grow equally well without it, so that the exclusion of air, which, of course, contains oxygen, is not always a protection, if one of the anaerobic bacteria, as the kinds are called which do not require oxygen, is sealed in the can.

Spoiling of food is caused by the development of bacteria or yeasts. Certain chemical changes are produced as shown by gases, odors, and flavors.

Bacteria grow luxuriantly in foods containing a good deal of nitrogenous material, if warmth and moisture are present. Among foods rich in nitrogenous substances are all kinds of meat, fish, eggs, peas, beans, lentils, milk, etc. These foods are difficult to preserve on account of the omnipresent bacteria. This is seen in warm, muggy weather, when fresh meat, fish, soups, milk, etc., spoil quickly. Bacteria do not develop in substances containing a large percentage of sugar, but they grow rapidly in a suitable wet substance which con [7] tains a small percentage of sugar. Yeasts grow very readily in dilute solutions containing sugars in addition to some nitrogenous and mineral matters. Fruits are usually slightly acid and in general do not support bacterial growth, and so it comes about that canned fruits are more commonly fermented by yeasts than by bacteria.

Some vegetable foods have so much acid and so little nitrogenous substance that very few bacteria or yeasts attack them. Lemons, cranberries, and rhubarb belong to this class.

Temperature is an important factor in the growth of bacteria and yeasts. There are many kinds of these organisms, and each kind grows best at a certain temperature, some at a very low one and others at one as high as 125° F., or more. However, most kinds of bacteria are destroyed if exposed for ten or fifteen minutes to the temperature of boiling water (212° F.); but, if the bacteria are spore producers, cooking must be continued for an hour or more to insure their complete destruction. Generally speaking, in order to kill the spores the temperature must be higher than that of boiling water, or the article to be preserved must be cooked for about two hours at a temperature of 212° F., or a shorter time at a higher temperature under pressure. Yeasts and their spores are, however, more easily destroyed by heat than bacteria spores. Hence, fruits containing little nitrogenous material are more easily protected from fermentation than nitrogenous foods in which in general fermentation is caused by bacteria. Of course, it is not possible to know what kinds of organisms are in the food one is about to can or bottle; but we do know that most fruits are not favorable to the growth of bacteria, and, as a rule, the yeasts which grow in fruits and fruit juice can be destroyed by cooking ten or fifteen minutes at a temperature of 212° F. If no living organisms are left, and the sterilization of all appliances has been thorough, there is no reason why the fruit, if properly sealed, should not keep, with but slight change of texture or flavor, for a year or longer, although canned fruits undergo gradual change and deterioration even under the most favorable conditions.

When fruit is preserved with a large amount of sugar (a pound of sugar to a pound of fruit) it does not need to be hermetically sealed to protect it from bacteria and yeasts, because the thick, sugary sirup formed is not favorable to their growth. However, the self-sealing jars are much better than keeping such fruit in large receptacles, from which it is taken as needed, because molds grow freely on moist, sugary substances exposed to the air.

MOLDS AND MOLDING.

Every housekeeper is familiar with molds which, under favorable conditions of warmth and moisture, grow upon almost any kind of [8] organic material. This is seen in damp, warm weather, when molds form in a short time on all sorts of starchy foods, such as boiled potatoes, bread, mush, etc., as well as fresh, canned, and preserved fruits.

Molds develop from spores which are always floating about in the air. When a spore falls upon a substance containing moisture and suitable food it sends out a fine thread, which branches and works its way over and into the attacked substance. In a short time spores are produced and the work of reproduction goes on.

In the first stages molds are white or light gray and hardly noticeable; but when spores develop the growth gradually becomes colored. In fact, the conditions of advanced growth might be likened to those of a flower garden. The threads — mycelium — might be likened to the roots of plants and the spores to the flower and seeds.

Mold spores are very light and are blown about by the wind. They are a little heavier than air, and drop on shelves, tables, and floor, and are easily set in motion again by the movement of a brush, duster, etc. If one of these spores drops on a jar of preserves or a tumbler of jelly, it will germinate if there be warmth and moisture enough in the storeroom. Molds do not ordinarily cause fermentation of canned foods, although they are the common cause of the decay of raw fruits. They are not as injurious to canned goods as are bacteria and yeasts. They do not penetrate deeply into preserves or jellies, or into liquids or semiliquids, but if given time they will, at ordinary room temperature, work all through suitable solid substances which contain moisture. Nearly every housekeeper has seen this in the molding of a loaf of bread or cake.

In the work of canning, preserving, and jelly making it is important that the food shall be protected from the growth of molds as well as the growth of yeasts and bacteria.

To kill mold spores food must be exposed to a temperature of from 150° F. to 212° F. After this it should be kept in a cool, dry place and covered carefully that no floating spore can find lodgment on its surface.

STERILIZATION.

To sterilize a substance or thing is to destroy all life and sources of life in and about it. In following the brief outline of the structure and work of bacteria, yeasts, and molds, it has been seen that damage to foods comes through the growth of these organisms on or in the food; also that if such organisms are exposed to a temperature of 212° F., life will be destroyed, but that spores and a few resisting bacteria are not destroyed at a temperature of 212° F., unless exposed to it for two or more hours.

Bacteria and yeasts, which are intimately mixed with food, are not [9] as easily destroyed as are those on smooth surfaces, such as the utensils and jars employed in the preparation of the food.

Since air and water, as well as the foods, contain bacteria and yeasts, and may contain mold spores, all utensils used in the process of preserving foods are liable to be contaminated with these organisms. For this reason all appliances, as well as the food, must be sterilized.

Stewpans, spoons, strainers, etc., may be put on the fire in cold or boiling water and boiled ten or fifteen minutes. Tumblers, bottles, glass jars, and covers should be put in cold water and heated gradually to the boiling point, and then boiled for ten or fifteen minutes. The jars must be taken one at a time from the boiling water at the moment they are to be filled with the boiling food. The work should be done in a well swept and dusted room, and the clothing of the workers and the towels used should be clean. The food to be sterilized should be perfectly sound and clean.

As in this bulletin we have only to do with fruits, it will not be necessary to say anything more about long cooking at a high temperature.

In canning fruits it is well to remember that the product is more satisfactory if heated gradually to the boiling point and then cooked the given time.

UTENSILS NEEDED FOR CANNING AND PRESERVING.

In preserving, canning, and jelly making iron or tin utensils should never be used. The fruit acids attack these metals and so give a bad color and metallic taste to the products. The preserving kettles should be porcelain lined, enameled, or of a metal that will not form troublesome chemical combinations with fruit juices. The kettles should be broad rather than deep, as the fruit should not be cooked in deep layers. Nearly all the necessary utensils may be found in some ware not subject to chemical action. A list of the most essential articles follows:

Two preserving kettles, 1 colander, 1 fine strainer, 1 skimmer, 1 ladle, 1 large-mouthed funnel, 1 wire frying basket, 1 wire sieve, 4 long-handled wooden spoons, 1 wooden masher, a few large pans, knives for paring fruit (plated if possible), flat-bottomed clothes boiler, wooden or willow rack to put in the bottom of the boiler, iron tripod or ring, squares of cheese cloth. In addition, it would be well to have a flannel straining bag, a frame on which to hang the bag, a sirup gauge and a glass cylinder, a fruit pricker, and plenty of clean towels.

The regular kitchen pans will answer for holding and washing the fruit. Mixing bowls and stone crocks can be used for holding the fruit juice and pared fruit. When fruit is to be plunged into boiling [10] water for a few minutes before paring, the ordinary stewpans may be employed for this purpose.

Fig. 1. — Wire basket.

Scales are a desirable article in every kitchen, as weighing is much more accurate than the ordinary measuring. But, knowing that a large percentage of the housekeepers do not possess scales, it has seemed wise to give all the rules in measure rather than weight.

If canning is done by the oven process, a large sheet of asbestos, for the bottom of the oven, will prevent the cracking of jars.

The wooden rack, on which the bottles rest in the washboiler, is made in this manner: Have two strips of wood measuring 1 inch high, 1 inch wide, and 2 inches shorter than the length of the boiler. On these pieces of wood tack thin strips of wood that are 1½ inches shorter than the width of the boiler. These cross-strips should be about 1 inch wide, and there should be an inch between two strips. This rack will support the jars and will admit the free circulation of boiling water about them. Young willow branches, woven into a mat, also make a good bed for bottles and jars.

Fig. 2. — Wire sieve.

The wire basket is a saver of time and strength (fig. 1). The fruit to be peeled is put into the basket, which is lowered into a deep kettle partially filled with boiling water. After a few minutes the basket is lifted from the boiling water, plunged for a moment into cold water, and the fruit is ready to have the skin drawn off.

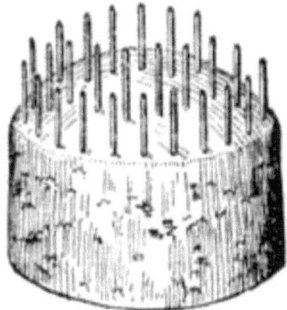

Fig. 3.—Fruit pricker.

A strong wire sieve is a necessity when purées of fruit are to be made (fig. 2). These sieves are known as purée sieves. They are made of strong wire and in addition have supports of still stronger wire.

A fruit pricker is easily made and saves time (fig. 3). Cut a piece half an inch deep from a broad cork; press through this a dozen or more coarse darning needles; tack the cork on a piece of board. Strike the [11] fruit on the bed of needles, and you have a dozen holes at once. When the work is finished, remove the cork from the board, wash and dry thoroughly. A little oil on the needles will prevent rusting. With needles of the size suggested there is little danger of the points breaking, but it is worth remembering that the use of pricking machines was abandoned in curing prunes on a commercial scale in California because the steel needles broke and remained in the fruit.

Fig. 4.—Wooden vegetable masher.

A wooden vegetable masher is indispensable when making jellies and purées (fig. 4).

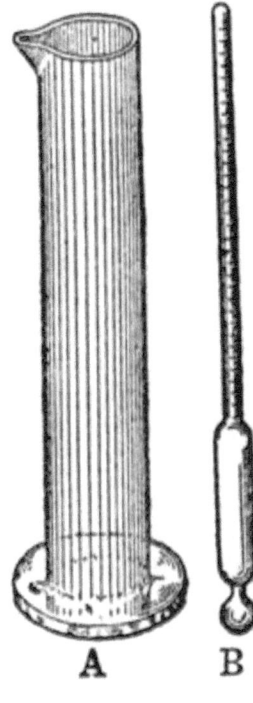

Fig. 5.—Glass cylinder (A) and sirup gauge (B).

A sirup gauge and glass cylinder (fig. 5 A and B) are not essential to preserving, canning, and jelly making, but they are valuable aids in getting the right proportion of sugar for fruit or jelly. The sirup gauge costs about 50 cents and the cylinder about 25 cents. A lipped cylinder that holds a little over a gill is the best size.

Small iron rings, such as sometimes come off the hub of cart wheels, may be used instead of a tripod for slightly raising the preserving kettles from the hot stove or range.

To make a flannel straining bag, take a square piece of flannel (27 by 27 inches is a good size), fold it to make a three-cornered bag, stitch one of the sides, cut the top square across, bind the opening with strong, broad tape, stitch on this binding four tapes with which to tie the bag to a frame.

To use this bag, tie it to a strong frame or to the backs of two kitchen chairs. If the chairs are used, place some heavy articles in them; or the bag may hang on a pole (a broom handle) which rests on the backs of the chairs. A high stool turned upside down makes a good support for the bag. Put a bowl on the floor under the bag, then pour in the fruit juice, which will pass through comparatively clear.

Before it is used the bag should be washed and boiled in clear water.

[12]

SELECTION AND PREPARATION OF THE FRUIT.

The selection of fruit is one of the first steps in obtaining successful results. The flavor of fruit is not developed until it is fully ripe, but the time at which the fruit is at its best for canning, jelly making, etc., is just before it is perfectly ripe. In all soft fruits the fermentative stage follows closely upon the perfectly ripe stage; therefore it is better to use underripe rather than overripe fruit. This is especially important in jelly making for another reason also: In overripe fruit the pectin begins to lose its jelly-making quality.

All fruits should, if possible, be freshly picked for preserving, canning, and jelly making. No imperfect fruit should be canned or preserved. Gnarly fruit may be used for jellies or marmalades by cutting out defective portions. Bruised spots should be cut out of peaches and pears. In selecting small-seeded fruits, like berries, for canning, those having a small proportion of seed to pulp should be chosen. In dry seasons berries have a larger proportion of seeds to pulp than in a wet or normal season, and it is not wise to can or preserve such fruit unless the seeds are removed. The fruit should be rubbed through a sieve that is fine enough to keep back the seeds. The strained pulp can be preserved as a purée or marmalade.

When fruit is brought into the house put it where it will keep cool and crisp until you are ready to use it.

The preparation of fruit for the various processes of preserving is the second important step. System will do much to lighten the work.

Begin by having the kitchen swept and dusted thoroughly, that there need not be a large number of mold spores floating about. Dust with a damp cloth. Have plenty of hot water and pans in which jars and utensils may be sterilized. Have at hand all necessary utensils, towels, sugar, etc.

Prepare only as much fruit as can be cooked while it still retains its color and crispness. Before beginning to pare fruit have some sirup ready, if that is to be used, or if sugar is to be added to the fruit have it weighed or measured.

Decide upon the amount of fruit you will cook at one time, then have two bowls—one for the sugar and one for the fruit—that will hold just the quantity of each. As the fruit is pared or hulled, as the case may be, drop it into its measuring bowl. When the measure is full put the fruit and sugar in the preserving kettle. While this is cooking another measure may be prepared and put in the second preserving kettle. In this way the fruit is cooked quickly and put in the jars and sealed at once, leaving the pans ready to sterilize another set of jars.

If the fruit is to be preserved or canned with sirup, it may be put [13] into the jars as fast as it is prepared. As soon as a jar is full, pour in enough sirup to cover it.

If several people are helping and large kettles are being used for the preserving, or where fruit (like quinces and hard pears) must be first boiled in clear water, the pared fruit should be dropped into a bowl of cold water made slightly acid with lemon juice (one tablespoonful of lemon juice to a quart of water). This will keep the fruit white.

All large, hard fruit must be washed before paring. Quinces should be rubbed with a coarse towel before they are washed.

If berries must be washed, do the work before stemming or hulling them. The best way to wash berries is to put a small quantity into a colander and pour cold water over them; then turn them on a sieve to drain. All this work must be done quickly that the fruit may not absorb much water.

Do not use the fingers for hulling strawberries. A simple huller can be bought for five cents.

If practicable pare fruit with a silver knife, so as not to stain or darken the product. The quickest and easiest way to peel peaches is to drop them into boiling water for a few minutes. Have a deep kettle a little more than half full of boiling water; fill a wire basket with peaches; put a long-handled spoon under the handle of the basket and lower into the boiling water. At the end of three minutes lift the basket out by slipping the spoon under the handle. Plunge the basket for a moment into a pan of cold water. Let the peaches drain a minute, then peel. Plums and tomatoes may be peeled in the same manner.

If the peaches are to be canned in sirup, put them at once into the sterilized jars. They may be canned whole or in halves. If in halves, remove nearly all the stones or pits. For the sake of the flavor, a few stones should be put in each jar.

When preparing cherries, plums, or crab apples for canning or preserving, the stem or a part of it may be left on the fruit.

When preparing to make jelly have ready the cheese-cloth strainer, enameled colander, wooden spoons, vegetable masher, measures, tumblers, preserving kettles, and sugar.

If currant jelly is to be made, free the fruit from leaves and large stems. If the jelly is to be made from any of the other small fruits, the stems and hulls must be removed.

When the jelly is to be made from any of the large fruits the important part of the preparation is to have the fruit washed clean, then to remove the stem and the blossom end. Nearly all the large fruits are better for having the skin left on. Apples and pears need not be cored. There is so much gummy substance in the cores of quinces that it is best not to use this portion in making fine jelly.

[14]

MAKING SIRUP FOR USE IN CANNING AND PRESERVING.

Such sirups as are used in canning and preserving are made with varying proportions of water and sugar. When the proportion of sugar is large and that of the water small the sirup is said to be heavy. When the water predominates the sirup is light.

There are several methods of measuring the proportion of sugar in a sirup. The most scientific and accurate is with the sirup gauge. Careful measurement or weighing is, however, quite satisfactory for all ordinary work if the sirup need not be boiled a long time. In boiling the water evaporates and the sirup grows thicker and richer. The amount of evaporation depends upon the surface exposed and the pressure of the atmosphere. For example, if a large quantity of sirup is boiled in a deep kettle the evaporation will not be rapid. If the same quantity of sirup were boiled the same length of time in a broad, shallow kettle the water would evaporate more rapidly and the sirup would be thicker and heavier. If a given quantity of sirup were boiled the same length of time in a high altitude, Colorado for example, and at the sea level, it would be found that the sirup boiled at the sea level would be thicker and less in volume than that boiled in Colorado. From this it will be seen that it is difficult to say what proportion of sugar a sirup will contain after it has been boiling ten or more minutes. Of course by the use of the sirup gauge the proportion of sugar in a sirup may be ascertained at any stage of the boiling. After all, however, it is possible to measure sugar and water so that you can know the percentage of sugar when the sirup begins to boil. The following statement gives the percentage of sugar at the time when the sirup has been boiling one minute and also what kind of sirup is suitable for the various kinds of fruit:

One pint sugar and 1 gill of water gives sirup of 40° density: Use for preserved strawberries and cherries.

One pint sugar and one-half pint water gives sirup of 32° density.

One pint sugar and 3 gills water gives sirup of 28° density: Use either this or the preceding for preserved peaches, plums, quinces, currants, etc.

One pint sugar and 1 pint water gives sirup of 24° density: Use for canned acid fruits.

One pint sugar and 1½ pints water gives sirup of 17° density.

One pint sugar and 2 pints water gives sirup of 14° density: Use either of these two light sirups for canned pears, peaches, sweet plums, and cherries, raspberries, blueberries, and blackberries.

The lightest sirups may be used for filling up the jars after they are taken from the oven or boiler. The process of making a sirup is very simple, but there are a few points that must be observed if sirup and fruit are to be perfect. Put the sugar and water in the saucepan and stir on the stove until all the sugar is dissolved. Heat slowly to the boiling point and boil gently without stirring. The length of time [15] that the sirup should boil will depend upon how rich it is to be. All sirups are better for boiling from ten to thirty minutes. If rich sirups are boiled hard, jarred, or stirred they are apt to crystallize. The sirup may be made a day or two in advance of canning time. The light sirups will not keep long unless sealed, but the heavy sirups keep well if covered well.

USE OF THE SIRUP GAUGE.

The sirup gauge is a graduated glass tube, with a weighted bulb, that registers from 0° to 50°, and that is employed to determine the quantity of sugar contained in a sirup.

If this gauge is placed in pure water the bulb will rest on the bottom of the cylinder or other container. If sugar be dissolved in the water the gauge will begin to float. The more sugar there is dissolved in the water the higher the gauge will rise. In making tests it is essential that the sirup should be deep enough to reach the zero point of the gauge. If a glass cylinder holding about half a gill is filled to about two-thirds its height, and the gauge is then placed in the cylinder, the quantity of sugar in the sirup will be registered on the gauge.

Experiments have demonstrated that when sugar is dissolved and heated in fruit juice, if the sirup gauge registers 25°, the proportion of sugar is exactly right for combining with the pectin bodies to make jelly. The sirup gauge and the glass cylinder must both be heated gradually that the hot sirup may not break them. If the gauge registers more than 25°, add a little more fruit juice. If, on the other hand, it registers less than 25°, add more sugar. In making sirups for canning and preserving fruits, the exact amount of sugar in a sirup may be ascertained at any stage of boiling, and the sirup be made heavier by adding sugar, or lighter by adding water, as the case demands.

CANNING FRUIT.

This method of preserving fruit for home use is from all points the most desirable. It is the easiest and commonly considered the most economical and the best, because the fruit is kept in a soft and juicy condition in which it is believed to be easily digested. The wise housekeeper will can her principal fruit supply, making only enough rich preserves to serve for variety and for special occasions.

The success of canning depends upon absolute sterilization. If the proper care is exercised there need be no failure, except in rare cases, when a spore has developed in the can. There are several methods of canning; and while the principle is the same in all methods, the conditions under which the housekeeper must do her work may, in her case, make one method more convenient than another. For this reason three will be given which are considered the best and easiest. These [16] are: Cooking the fruit in the jars in an oven; cooking the fruit in the jars in boiling water; and stewing the fruit before it is put in the jars. The quantity of sugar may be increased if the fruit is liked sweet.

It is most important that the jars, covers, and rubber rings be in perfect condition. Examine each jar and cover to see that there is no defect in it. Use only fresh rubber rings, for if the rubber is not soft and elastic the sealing will not be perfect. Each year numbers of jars of fruit are lost because of the false economy in using an old ring that has lost its softness and elasticity. Having the jars, covers, and rings in perfect condition, the next thing is to wash and sterilize them.

Have two pans partially filled with cold water. Put some jars in one, laying them on their sides, and some covers in the other. Place the pans on the stove where the water will heat to the boiling point. The water should boil at least ten or fifteen minutes. Have on the stove a shallow milk pan in which there is about 2 inches of boiling water. Sterilize the cups, spoons, and funnel, if you use one, by immersing in boiling water for a few minutes. When ready to put the prepared fruit in the jars slip a broad skimmer under a jar and lift it and drain free of water. Set the jar in the shallow milk pan and fill to overflowing with the boiling fruit. Slip a silver-plated knife or the handle of a spoon around the inside of the jar, that the fruit and juice may be packed solidly. Wipe the rim of the jar, dip the rubber ring in boiling water and put it smoothly on the jar, then put on the cover and fasten. Place the jar on a board and out of a draft of cold air. The work of filling and sealing must be done rapidly, and the fruit must be boiling hot when it is put into the jars. If screw covers are used, it will be necessary to tighten them after the glass has cooled and contracted. When the fruit is cold wipe the jars with a wet cloth. Paste on the labels, if any, and put the jars on shelves in a cool, dark closet.

In canning, any proportion of sugar may be used, or fruit may be canned without the addition of any sugar. However, that which is designed to be served as a sauce should have the sugar cooked with it. Fruit intended for cooking purposes need not have the sugar added to it.

Juicy fruits, such as berries and cherries, require little or no water. Strawberries are better not to have water added to them. The only exception to this is when they are cooked in a heavy sirup.

RASPBERRIES.

- 12 quarts of raspberries.
- 2 quarts of sugar.

Put 2 quarts of the fruit in the preserving kettle; heat slowly on the stove; crush with a wooden vegetable masher; spread a square of [17] cheese cloth over a bowl, and turn the crushed berries and juice into it. Press out the juice, which turn into the preserving kettle. Add the sugar and put on the stove; stir until the sugar is dissolved. When the sirup begins to boil, add the remaining 10 quarts of berries. Let them heat slowly. Boil ten minutes, counting from the time they begin to bubble. Skim well while boiling. Put in cans and seal as directed.

RASPBERRIES AND CURRANTS.

- 10 quarts of raspberries.

- 3 quarts of currants.
- 2½ quarts of sugar.

Heat, crush, and press the juice from the currants and proceed as directed for raspberries.

BLACKBERRIES.

The same as for raspberries.

CURRANTS.

- 12 quarts of currants.
- 4 quarts of sugar.

Treat the same as for raspberries.

GOOSEBERRIES.

- 6 quarts of berries.
- 1½ quarts of sugar.
- 1 pint of water.

For green gooseberries dissolve the sugar in the water, then add the fruit and cook fifteen minutes. Ripe gooseberries are to be treated the same as the green fruit, but use only half as much water. Green gooseberries may also be canned the same as rhubarb (see p. 18).

BLUEBERRIES.

- 12 quarts of berries.
- 1 quart of sugar.
- 1 pint of water.

Put water, berries, and sugar in the preserving kettle; heat slowly. Boil fifteen minutes, counting from the time the contents of the kettle begin to bubble.

CHERRIES.

- 6 quarts of cherries.
- 1½ quarts of sugar.
- ½ pint of water.

Measure the cherries after the stems have been removed. Stone them or not, as you please. If you stone them be careful to save all the juice. Put the sugar and water in the preserving kettle and stir [18] over the fire until the sugar is dissolved. Put in the cherries and heat slowly to the boiling point. Boil ten minutes, skimming carefully.

GRAPES.

- 6 quarts of grapes.
- 1 quart of sugar.
- 1 gill of water.

Squeeze the pulp of the grapes out of the skins. Cook the pulp five minutes and then rub through a sieve that is fine enough to hold back the seeds. Put the water, skins, and pulp into the preserving kettle and heat slowly to the boiling point. Skim the fruit and then add the sugar. Boil fifteen minutes.

Sweet grapes may be canned with less sugar; very sour ones may have more.

RHUBARB.

Cut the rhubarb when it is young and tender. Wash it thoroughly and then pare; cut into pieces about 2 inches long. Pack in sterilized jars. Fill the jars to overflowing with cold water and let them stand ten minutes. Drain off the water and fill again to overflowing with fresh cold water. Seal with sterilized rings and covers. When required for use, treat the same as fresh rhubarb.

Green gooseberries may be canned in the same manner. Rhubarb may be cooked and canned with sugar in the same manner as gooseberries.

PEACHES.

- 8 quarts of peaches.
- 1 quart of sugar.
- 3 quarts of water.

Put the sugar and water together and stir over the fire until the sugar is dissolved. When the sirup boils skim it. Draw the kettle back where the sirup will keep hot but not boil.

Pare the peaches, cut in halves, and remove the stones, unless you prefer to can the fruit whole.

Put a layer of the prepared fruit into the preserving kettle and cover with some of the hot sirup. When the fruit begins to boil, skim carefully. Boil gently for ten minutes, then put in the jars and seal. If the fruit is not fully ripe it may require a little longer time to cook. It should be so tender that it may be pierced easily with a silver fork. It is best to put only one layer of fruit in the preserving kettle. While this is cooking the fruit for the next batch may be pared.

PEARS.

If the fruit is ripe it may be treated exactly the same as peaches. If, on the other hand, it is rather hard it must be cooked until so tender that a silver fork will pierce it readily.

[19]

QUINCES.

- 4 quarts of pared, cored, and quartered quinces.
- 1½ quarts of sugar.

- 2 quarts of water.

Rub the fruit hard with a coarse, crash towel, then wash and drain. Pare, quarter, and core; drop the pieces into cold water (see p. 13). Put the fruit in the preserving kettle with cold water to cover it generously. Heat slowly and simmer gently until tender. The pieces will not all require the same time to cook. Take each piece up as soon as it is so tender that a silver fork will pierce it readily. Drain on a platter. Strain the water in which the fruit was cooked through cheese cloth. Put two quarts of the strained liquid and the sugar into the preserving kettle; stir over the fire until the sugar is dissolved. When it boils skim well and put in the cooked fruit. Boil gently for about twenty minutes.

CRAB APPLES.

- 6 quarts of apples.
- 1½ quarts of sugar.
- 2 quarts of water.

Put the sugar and water into the preserving kettle. Stir over the fire until the sugar is dissolved. When the sirup boils skim it.

Wash the fruit, rubbing the blossom end well. Put it in the boiling sirup, and cook gently until tender. It will take from twenty to fifty minutes, depending upon the kind of crab apples.

PLUMS.

- 8 quarts of plums.
- 2 quarts of sugar.
- 1 pint of water.

Nearly all kinds of plums can be cooked with the skins on. If it is desired to remove the skin of any variety, plunge them in boiling water for a few minutes. When the skins are left on, prick them thoroughly to prevent bursting. (See fruit pricker, p. 10.)

Put the sugar and water into the preserving kettle and stir over the fire until the sugar is dissolved. Wash and drain the plums. Put some of the fruit in the boiling sirup. Do not crowd it. Cook five minutes; fill and seal the jars. Put more fruit in the sirup. Continue in this manner until all the fruit is done. It may be that there will not be sufficient sirup toward the latter part of the work; for this reason it is well to have a little extra sirup on the back of the stove.

STEWED TOMATOES.

Wash the tomatoes and plunge into boiling water for five minutes. Pare and slice, and then put into the preserving kettle; set the kettle [20] on an iron ring. Heat the tomatoes slowly, stirring frequently from the bottom. Boil for thirty minutes, counting from the time the vegetable begins actually to boil. Put in sterilized jars and seal.

WHOLE TOMATOES.

- 8 quarts of medium-sized tomatoes.
- 4 quarts of sliced tomatoes.

Put the pared and sliced tomatoes into a stewpan and cook as directed for stewed tomatoes. When they have been boiling twenty minutes take from the fire and rub through a strainer. Return to the fire.

While the sliced tomatoes are cooking, pare the whole tomatoes and put them in sterilized jars. Pour into the jars enough of the stewed and strained tomato to fill all the interstices. Put the uncovered jars in a moderate oven, placing them on a pad of asbestos or in shallow pans of hot water. Let the vegetable cook in the oven for half an hour. Take from the oven and fill to overflowing with boiling hot, strained tomato, then seal. If there is any of the strained tomato left, can it for sauces.

CANNED FRUIT COOKED IN THE OVEN.

This method of canning fruit, in the opinion of the writer, is the one to be preferred. The work is easily and quickly done, and the fruit retains its shape, color, and flavor better than when cooked in the preserving kettle.

Cover the bottom of the oven with a sheet of asbestos, the kind plumbers employ in covering pipes. It is very cheap and may usually be found at plumbers' shops. If the asbestos is not available, put into the oven shallow pans in which there are about two inches of boiling water.

Sterilize the jars and utensils. Make the sirup; prepare the fruit the same as for cooking in the preserving kettle. Fill the hot jars with it, and pour in enough sirup to fill the jar solidly. Run the blade of a silver-plated knife around the inside of the jar. Place the jars in the oven, either on the asbestos or in the pan of water. The oven should be moderately hot. Cook the fruit ten minutes; remove from the oven and fill the jar with boiling sirup. Wipe and seal. Place the jars on a board and out of a draft of air. If the screw covers are used tighten them after the glass has cooled.

Large fruits, such as peaches, pears, quinces, crab apples, etc., will require about a pint of sirup to each quart jar of fruit. The small fruit will require a little over half a pint of sirup.

The amount of sugar in each quart of sirup should be regulated to suit the fruit with which it is to be used. The data on page 14 will be a guide. The quantities given will not make the fruit very sweet. The quantity of sugar may be increased or diminished to suit the taste.

[21]

CANNED FRUIT COOKED IN A WATER BATH.

Prepare the fruit and sirup as for cooking in the oven.

Fill the sterilized jars and put the covers on loosely. Have a wooden rack in the bottom of a wash boiler (see p. 10). Put in enough warm water to come to about 4 inches above the rack. Place the filled jars in the boiler, but do not let them touch one another. Pack clean white cotton rags, or perhaps better, cotton rope, between and around the jars to prevent them from striking one another when the water begins to boil. Cover the boiler and let the fruit cook ten minutes from the time the water surrounding it begins to boil.

Draw the boiler back and take off the cover. When the steam passes off take out one jar at a time and place in a pan of boiling water beside the boiler, fill up with boiling sirup, and seal. Put the jars on a board and do not let cold air blow upon them. If screw covers are used tighten them when the glass has cooled and contracted.

PRESERVING FRUIT.

In the case of most fruits, canning with a little sugar is to be preferred to preserving with a large quantity of sugar. There are, however, some fruits that are only good when preserved with a good deal of sugar. Of course, such preparations of fruit are only desirable for occasional use. The fruits best adapted for preserving are strawberries, sour cherries, sour plums, and quinces. Such rich preparations should be put up in small jars or tumblers.

STRAWBERRIES.

Use equal weights of sugar and strawberries. Put the strawberries in the preserving kettle in layers, sprinkling sugar over each layer. The fruit and sugar should not be more than 4 inches deep. Place the kettle on the stove and heat the fruit and sugar slowly to the boiling point. When it begins to boil skim carefully. Boil ten minutes, counting from the time the fruit begins to bubble. Pour the cooked fruit into platters, having it about 2 or 3 inches deep. Place the platters in a sunny window, in an unused room, for three or four days. In that time the fruit will grow plump and firm, and the sirup will thicken almost to a jelly. Put this preserve, cold, into jars or tumblers.

WHITE CURRANTS.

Select large, firm fruit, remove the stems, and proceed as for strawberries.

CHERRIES.

The sour cherries, such as Early Richmond and Montmorency, are best for this preserve. Remove the stems and stones from the cherries and proceed as for strawberry preserve.

[22]

CHERRIES PRESERVED WITH CURRANT JUICE.

- 12 quarts of cherries.
- 3 quarts of currants.
- 2 quarts of sugar.

Put the currants in the preserving kettle and on the fire. When they boil up crush them and strain through cheese cloth, pressing out all the juice.

Stem and stone the cherries, being careful to save all the juice. Put the cherries, fruit juice, and sugar in the preserving kettle. Heat to the boiling point and skim carefully. Boil for twenty minutes. Put in sterilized jars or tumblers. This gives an acid preserve. The sugar may be doubled if richer preserves are desired.

PLUM PRESERVE.

- 4 quarts of green gages.
- 2 quarts of sugar.
- 1 pint of water.

Prick the fruit and put it in a preserving kettle. Cover generously with cold water. Heat to the boiling point and boil gently for five minutes. Drain well.

Put the sugar and water in a preserving kettle and stir over the fire until the sugar is dissolved. Boil five minutes, skimming well. Put the drained green gages in this sirup and cook gently for twenty minutes. Put in sterilized jars.

Other plums may be preserved in the same manner. The skins should be removed from white plums.

QUINCES.

- 4 quarts of pared, quartered, and cored quinces.
- 2 quarts of sugar.

- 1 quart of water.

Boil the fruit in clear water until it is tender, then skim out and drain.

Put the 2 quarts of sugar and 1 quart of water in the preserving kettle; stir until the sugar is dissolved. Let it heat slowly to the boiling point. Skim well and boil for twenty minutes. Pour one-half of the sirup into a second kettle. Put one-half of the cooked and drained fruit into each kettle. Simmer gently for half an hour, then put in sterilized jars. The water in which the fruit was boiled can be used with the parings, cores, and gnarly fruit to make jelly.

FRUIT PURÉES.

Purées of fruit are in the nature of marmalades, but they are not cooked so long, and so retain more of the natural flavor of the fruit. [23] This is a particularly nice way to preserve the small, seedy fruits, which are to be used in puddings, cake, and frozen desserts.

Free the fruit from leaves, stems, and decayed portions. Peaches and plums should have the skins and stones removed. Rub the fruit through a purée sieve. To each quart of the strained fruit add a pint of sugar. Pack in sterilized jars. Put the covers loosely on the jars. Place the jars on the rack in the boiler. Pour in enough cold water to come half way up the sides of the jars. Heat gradually to the boiling point and boil thirty minutes, counting from the time when the water begins to bubble.

Have some boiling sirup ready. As each jar is taken from the boiler put it in a pan of hot water and fill up with the hot sirup. Seal at once.

MARMALADES.

Marmalades require great care while cooking because no moisture is added to the fruit and sugar. If the marmalade is made from berries the fruit should be rubbed through a sieve to remove the seeds. If large fruit is used have it washed, pared, cored, and quartered.

Measure the fruit and sugar, allowing one pint of sugar to each quart of fruit.

Rinse the preserving kettle with cold water that there may be a slight coat of moisture on the sides and bottom. Put alternate layers of fruit and sugar in the kettle, having the first layer fruit. Heat slowly, stirring frequently. While stirring, break up the fruit as much as possible. Cook about two hours, then put in small sterilized jars.

FRUIT PRESERVED IN GRAPE JUICE.

Any kind of fruit can be preserved by this method, but it is particularly good for apples, pears, and sweet plums. No sugar need be used in this process.

Boil 6 quarts of grape juice in an open preserving kettle, until it is reduced to 4 quarts. Have the fruit washed and pared, and, if apples or pears, quartered and cored. Put the prepared fruit in a preserving kettle and cover generously with the boiled grape juice. Boil gently until the fruit is clear and tender, then put in sterilized jars.

BOILED CIDER.

When the apple crop is abundant and a large quantity of cider is made, the housekeeper will find it to her advantage to put up a generous supply of boiled cider. Such cider greatly improves mincemeat, and can be used at any time of the year to make cider apple sauce. It is also a good selling article.

The cider for boiling must be perfectly fresh and sweet. Put it in a large, open preserving kettle and boil until it is reduced one-half. [24] Skim frequently while boiling. Do not have the kettle more than two-thirds full.

Put in bottles or stone jugs.

CIDER APPLE SAUCE.

- 5 quarts of boiled cider.
- 8 quarts of pared, quartered, and cored sweet apples.

Put the fruit in a large preserving kettle and cover with the boiled cider. Cook slowly until the apples are clear and tender. To prevent burning, place the kettle on an iron tripod or ring. It will require from two to three hours to cook the apples. If you find it necessary to stir the sauce be careful to break the apples as little as possible. When the sauce is cooked, put in sterilized jars.

In the late spring, when cooking apples have lost much of their flavor and acidity, an appetizing sauce may be made by stewing them with diluted boiled cider, using 1 cupful of cider to 3 of water.

CIDER PEAR SAUCE.

Cooking pears may be preserved in boiled cider the same as sweet apples. If one prefers the sauce less sour, 1 pint of sugar may be added to each quart of boiled cider.

METHODS OF MAKING JELLY.

In no department of preserving does the housekeeper feel less sure of the result than in jelly making. The rule that works perfectly one time fails another time. Why this is so the average housekeeper does not know; so there is nearly always an element of uncertainty as to the result of the work. These two questions are being constantly asked: "Why does not my jelly harden?" "What causes my jelly to candy?"

It is an easy matter to say that there is something in the condition of the fruit, or that the fruit juice and sugar were cooked too short or too long a time. These explanations are often true; but they do not help the inquirer, since at other times just that proportion of sugar and time of cooking have given perfect jelly. In the following pages an attempt is made to give a clear explanation of the principles underlying the process of jelly making. It is believed that the women who study this carefully will find the key to unvarying success in this branch of preserving.

PECTIN, PECTOSE, PECTASE.

In all fruits, when ripe or nearly so, there is found pectin, a carbohydrate somewhat similar in its properties to starch. It is because of this substance in the fruit juice that we are able to make jelly. When equal quantities of sugar and fruit juice are combined and the mixture [25] is heated to the boiling point for a short time, the pectin in the fruit gelatinizes the mass.

It is important that the jelly maker should understand when this gelatinizing agent is at its best. Pectose and pectase always exist in the unripe fruit. As the fruit ripens the pectase acts upon the pectose, which is insoluble in water, converting it into pectin, which is soluble. Pectin is at its best when the fruit is just ripe or a little before. If the juice ferments, or the cooking of the jelly is continued too long, the pectin undergoes a change and loses its power of gelatinizing. It is, therefore, of the greatest importance that the fruit should be fresh, just ripe or a little underripe, and that the boiling of the sugar and juice should not be continued too long.

Fruits vary as to the quantities of sugar, acid, pectin, and gums in their composition. Some of the sour fruits contain more sugar than do some of the milder-flavored fruits. Currants, for example, often contain four or five times as much sugar as the peach. The peach does not contain so much free acid and it does contain a great deal of pectin bodies, which mask the acid; hence, the comparative sweetness of the ripe fruit.

SELECTION AND HANDLING OF FRUIT FOR JELLY MAKING.

An acid fruit is the most suitable for jelly making, though in some of the acid fruits, the strawberry, for example, the quantity of the jelly-making pectin is so small that it is difficult to make jelly with this fruit. If, however, some currant juice be added to the strawberry juice a pleasant jelly will be the result; yet, of course, the flavor of the strawberry will be modified. Here is a list of the most desirable fruits for jelly making. The very best are given first: Currant, crab apple, apple, quince, grape, blackberry, raspberry, peach.

Apples make a very mild jelly, and it may be flavored with fruits, flowers, or spices. If the apples are acid it is not advisable to use any flavor.

Juicy fruits, such as currants, raspberries, etc., should not be gathered after a rain, for they will have absorbed so much water as to make it difficult, without excessive boiling, to get the juice to jelly.

If berries are sandy or dusty it will be necessary to wash them, but the work should be done very quickly so that the fruit may not absorb much water. (See washing fruit, p. 13.)

Large fruits, such as apples, peaches, and pears, must be boiled in water until soft. The strained liquid will contain the flavoring matter and pectin.

It requires more work and skill to make jellies from the fruits to which water must be added than from the juicy fruits. If the juicy fruits are gathered at the proper time one may be nearly sure that they contain the right proportion of water. If gathered after a rain [26] the fruit must be boiled a little longer that the superfluous water may pass off in steam.

In the case of the large fruits a fair estimate is 3 quarts of strained juice from 8 quarts of fruit and about 4 quarts of water. If the quantity of juice is greater than this it should be boiled down to 3 quarts.

Apples will always require 4 quarts of water to 8 quarts of fruit, but juicy peaches and plums will require only 3 or 3½ quarts.

The jelly will be clearer and finer if the fruit is simmered gently and not stirred during the cooking.

It is always best to strain the juice first through cheese cloth and without pressure. If the cloth is double the juice will be quite clear. When a very clear jelly is desired the strained juice should pass through a flannel or felt bag. The juice may be pressed from the fruit left in the strainer and used in marmalade or for a second-quality jelly.

To make jelly that will not crystallize (candy) the right proportion of sugar must be added to the fruit juice. If the fruit contains a high percentage of sugar, the quantity of added sugar should be a little less than the quantity of fruit juice. That is to say, in a season when there has been a great deal of heat and sunshine there will be more sugar in the fruit than in a cold, wet season; consequently, 1 pint of currant juice will require but three-quarters of a pint of sugar. But in a cold, wet season the pint of sugar for the pint of juice must be measured generously.

Another cause of the jelly crystallizing is hard boiling. When the sirup boils so rapidly that particles of it are thrown on the upper part of the sides of the preserving kettle they often form crystals. If these crystals are stirred into the sirup they are apt to cause the mass to crystallize in time.

The use of the sirup gauge and care not to boil the sirup too violently would do away with all uncertainty in jelly making. The sirup gauge should register 25°, no matter what kind of fruit is used. (See p. 15.)

Jellies should be covered closely and kept in a cool, dry, dark place.

CURRANT JELLY.

The simplest method of making currant jelly is perhaps the following: Free the currants from leaves and large stems. Put them in the preserving kettle; crush a few with a wooden vegetable masher or spoon; heat slowly, stirring frequently.

When the currants are hot, crush them with the vegetable masher. Put a hair sieve or strainer over a large bowl; over this spread a double square of cheese cloth. Turn the crushed fruit and juice into the cheese cloth, and let it drain as long as it drips, but do not use [27] pressure. To hasten the process take the corners of the straining cloth firmly in the hands and lift from the sieve; move the contents by raising one side of the cloth and then the other. After this put the cloth over another bowl. Twist the ends together and press out as much juice as possible. This juice may be used to make a second quality of jelly.

The clear juice may be made into jelly at once, or it may be strained through a flannel bag. In any case, the method of making the jelly is the same.

Measure the juice, and put it in a clean preserving kettle. For every pint of juice add a pint of granulated sugar.

Stir until the sugar is dissolved, then place over the fire; watch closely, and when it boils up draw it back and skim; put over the fire again, and boil and skim once more; boil and skim a third time; then pour into hot glasses taken from the pan of water on the stove and set on a board. Place the board near a sunny window in a room where there is no dust. It is a great protection and advantage to have sheets of glass to lay on top of the tumblers. As soon as the jelly is set cover by one of the three methods given. (See p. 29.)

To make very transparent currant jelly, heat, crush, and strain the currants as directed in the simplest process. Put the strained juice in the flannel bag and let it drain through. Measure the juice and sugar, pint for pint, and finish as directed above.

To make currant jelly by the cold process follow the first rule for jelly as far as dissolving the sugar in the strained juice. Fill warm, sterilized glasses with this. Place the glasses on a board and put the board by a sunny window. Cover with sheets of glass and keep by the window until the jelly is set. The jelly will be more transparent if the juice is strained through the flannel bag. Jelly made by the cold process is more delicate than that made by boiling, but it does not keep quite so well.

RASPBERRY AND CURRANT JELLY.

Make the same as currant jelly, using half currants and half raspberries.

RASPBERRY JELLY.

Make the same as currant jelly.

BLACKBERRY JELLY.

Make the same as currant jelly.

STRAWBERRY JELLY.

To 10 quarts of strawberries add 2 quarts of currants and proceed as for currant jelly, but boil fifteen minutes.

[28]

RIPE-GRAPE JELLY.

An acid grape is best for this jelly. The sweet, ripe grapes contain too much sugar. Half-ripe fruit, or equal portions of nearly ripe and green grapes, will also be found satisfactory. Wild grapes make delicious jelly. Make the same as currant jelly.

GREEN-GRAPE JELLY.

Make the same as apple jelly.

PLUM JELLY.

Use an underripe acid plum. Wash the fruit and remove the stems. Put into the preserving kettle with 1 quart of water for each peck of fruit. Cook gently until the plums are boiled to pieces. Strain the juice and proceed the same as for currant jelly.

APPLE JELLY.

Wash, stem, and wipe the apples, being careful to clean the blossom end thoroughly. Cut into quarters and put into the preserving kettle. Barely cover with cold water (about 4 quarts of water to 8 of apples) and cook gently until the apples are soft and clear. Strain the juice and proceed as for currant jelly. There should be but 3 quarts of juice from 8 quarts of apples and 4 of water.

Apples vary in the percentage of sugar and acid they contain. A fine-flavored acid apple should be employed when possible. Apple jelly may be made at any time of the year, but winter apples are best and should be used when in their prime, i. e., from the fall to December or January. When it is found necessary to make apple jelly in the spring, add the juice of one lemon to every pint of apple juice.

CIDER APPLE JELLY.

Make the same as plain apple jelly, but covering the apples with cider instead of water. The cider must be fresh from the press.

CRAB-APPLE JELLY.

Make the same as plain apple jelly.

QUINCE JELLY.

Rub the quinces with a coarse crash towel; cut out the blossom end. Wash the fruit and pare it and cut in quarters. Cut out the cores, putting them in a dish by themselves. Have a large bowl half full of water; drop the perfect pieces of fruit into this bowl. Put the parings and imperfect parts, cut very fine, into the preserving kettle. Add a quart of water to every 2 quarts of fruit and parings. Put on the fire and cook gently for two hours. Strain and finish the same as [29] apple jelly. The perfect fruit may be preserved or canned.

To make quince jelly of a second quality, when the parings and fruit are put on to cook put the cores into another kettle and cover them generously with water and cook two hours. After all the juice has been drained from the parings and fruit, put what remains into the preserving kettle with the cores. Mix well and turn into the straining cloth. Press all the juice possible from this mixture. Put the juice in the preserving kettle with a pint of sugar to a pint of juice; boil ten minutes.

WILD FRUITS FOR JELLIES.

Wild raspberries, blackberries, barberries, grapes, and beach plums all make delicious jellies. The frequent failures in making barberry jelly come from the fruit not being fresh or from being overripe.

PREPARATION OF THE GLASSES FOR JELLY.

Sterilize the glasses; take from the boiling water and set them in a shallow baking pan in which there is about 2 inches of boiling water.

COVERING JELLIES.

Jellies are so rich in sugar that they are protected from bacteria and yeasts, but they must be covered carefully to protect them from mold spores and evaporation. The following methods of covering jellies are all good:

Have disks of thick white paper the size of the top of the glass. When the jelly is set, brush the top over with brandy or alcohol. Dip a disk of paper in the spirits and put it on the jelly. If the glasses have covers, put them on. If there are no covers, cut disks of paper about half an inch in diameter larger than the top of the glass. Beat together the white of one egg and a tablespoonful of cold water. Wet the paper covers with this mixture and put over the glass, pressing down the sides well to make them stick to the glass; or the covers may be dipped in olive oil and be tied on the glasses, but they must be cut a little larger than when the white of egg is used.

A thick coating of paraffin makes a good cover, but not quite so safe as the paper dipped in brandy or alcohol, because the spirits destroy any mold spores that may happen to rest on the jelly. If such spores are covered with the paraffin they may develop under it. However, the paper wet with spirits could be put on first and the paraffin poured over it.

If paraffin is used, break it into pieces and put in a cup. Set the cup in a pan of warm water on the back of the stove. In a few moments it will be melted enough to cover the jelly. Have the coating about a fourth of an inch thick. In cooling the paraffin contracts, and if the layer is very thin it will crack and leave a portion of the jelly exposed.

[30]

CANNED OR BOTTLED FRUIT JUICES.

Fruit juice is most desirable for drinking or for culinary purposes. Grape juice is particularly good as a drink. It may be canned with or without sugar but, except where the grapes have a large percentage of sugar, as is the case in California, some sugar should be added to the juice in canning.

Currant juice may be sterilized and canned without sugar. This juice may be made into jelly at any season of the year.

Fruit juices that are designed for use in frozen creams and water ices should be canned with a generous amount of sugar.

For grape juice good bottles are to be preferred to fruit cans. If you can get the self-sealing bottles, such as pop or beer comes in, the work of putting up grape juice will be light. If bottles are employed, be very careful to sterilize both bottles and corks.

GRAPE JUICE.

Wash the grapes and pick from the stems. Put the fruit in the preserving kettle and crush slightly. Heat slowly and boil gently for half an hour. Crush the fruit with a wooden spoon.

Put a sieve or colander over a large bowl and spread a square of cheese cloth over the sieve. Turn the fruit and juice into the cheese cloth; drain well, then draw the edges of the cheese cloth together and twist hard to press out all the juice possible.

Put the strained juice in a clean preserving kettle and on the fire. When it boils up, draw back and skim. Let it boil up again and skim; then add the sugar and stir until dissolved. Boil five minutes, skimming carefully. Fill hot sterilized jars or bottles. Put the jars or bottles in a moderate oven for ten minutes, in pans of boiling water. Have some boiling juice and pour a little of it into the jars as they are taken from the oven; then seal. Place on boards and set aside out of a cold draft.

A good proportion of sugar and juice is 1 gill of sugar to a quart of juice. The preparation and use of grape juice has been discussed at length in an earlier bulletin of this series. [a]

RASPBERRY, BLACKBERRY, STRAWBERRY, AND CURRANT JUICES.

With all these fruits except currants, proceed the same as for grape juice, but adding half a pint of sugar to each quart of juice. Currants will require 1 pint of sugar to a quart of juice.

CHERRY, PLUM, AND PEACH JUICES.

To preserve the juice of cherries, plums, peaches, and similar fruits, proceed as for jelly, but adding to each quart of juice half a pint of sugar instead of a quart as for jelly. If it is not desired to have the fruit juice transparent, the pulp of the fruit may be pressed to extract all the liquid.

[31]

FRUIT SIRUPS.

The only difference between sirups and juice is that in the sirup there must be at least half as much sugar as fruit juice.

These sirups are used for flavoring ice creams and water ices. They also make a delicious drink, when two or three spoonfuls are added to a glass of ice water.

RASPBERRY VINEGAR.

Put 4 quarts of raspberries in a bowl and pour over them 2 quarts of vinegar. Cover and set in a cool place for two days. On the second day strain the vinegar through cheese cloth. Put 4 quarts of fresh raspberries in the bowl and pour over them the vinegar strained from the first raspberries. Put in a cool place for two days, then strain. Put the strained juice in a preserving kettle with 3 quarts of sugar. Heat slowly, and when the vinegar boils skim carefully. Boil twenty minutes, then put in sterilized bottles.

About 2 tablespoonfuls of vinegar to a glass of water makes a refreshing drink.

Similar vinegars may be made from blackberries and strawberries.

FOOTNOTES:

[a] U. S. Dept. Agr., Farmers' Bulletin No. 175.

www.ingramcontent.com/pod-product-compliance
Lightning Source LLC
Chambersburg PA
CBHW030505220526
45464CB00006B/2664